Uwe H. Sueltz

PHILIPS 1962 - 2002

1. PHILIPS FROM 1963 TO 2002
2. A SELECTION FROM PHILIPS MANUFACTURING

Bibliografische Information durch die Deutsche Nationalbibliothek

Die Deutsche Nationalbibliothek verzeichnet diese Publikation in der Deutschen Nationalbibliografie; detaillierte bibliografische Daten sind im Internet über http://dnb.dnb.de abrufbar.

© 2025 Uwe H. Sültz, Verlag:

BoD · Books on Demand GmbH, In de Tarpen 42, 22848 Norderstedt, bod@bod.de

ISBN: 978-3-7693-2593-5

Druck:

Libri Plureos GmbH, Friedensallee 273, 22763 Hamburg

PHILIPS from 1963 to 2002

In this picture book I show PHILIPS Compact Cassettes from 1963 to 2002.

I also show the unknown one-hole cassette. PHILIPS never published it. The Compact Cassette, by Team Lou Ottens, was selected in 1962.

In diesem Bildband werden PHILIPS Cassetten von 1963 bis 2002 gezeigt.

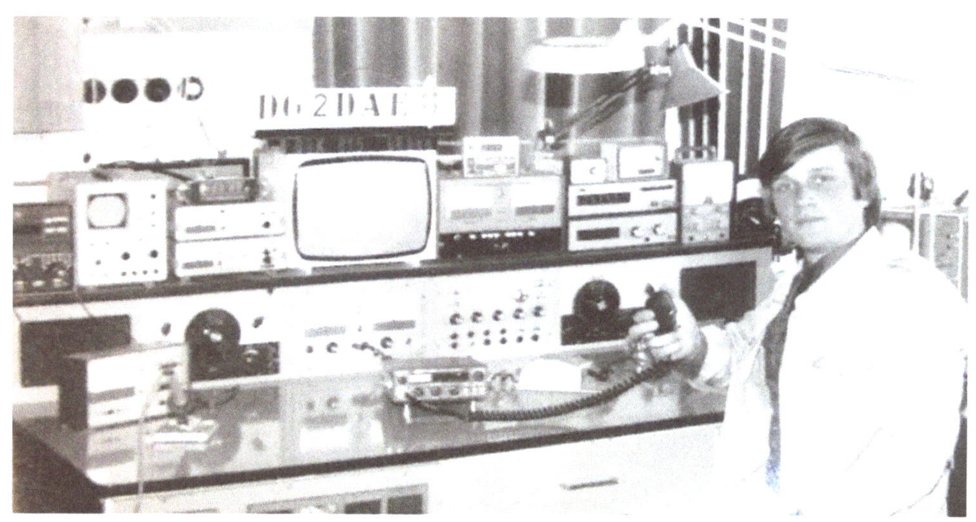

Uwe H. Sültz

Die erste Compact Cassette EL 1903 wurde am 8.1.1963 der Presse vorgestellt. Den Compact Cassetten Recorder EL 3300 zeigte PHILIPS dann auf der Funkausstellung 1963.

The first Compact Cassette EL 1903 was presented to the press on January 8, 1963. PHILIPS then showed the EL 3300 compact cassette recorder at the 1963 radio exhibition.

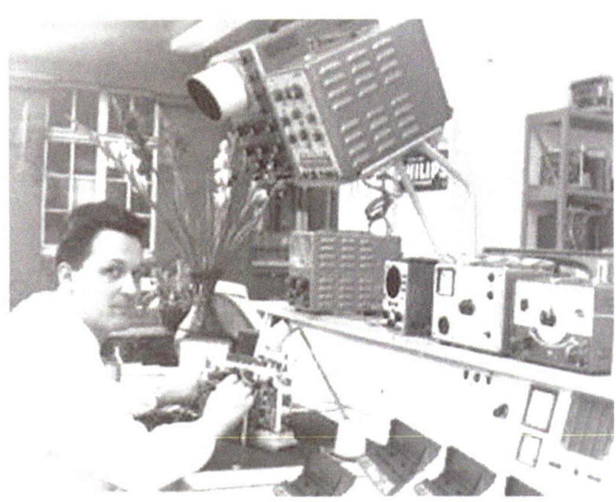

1963

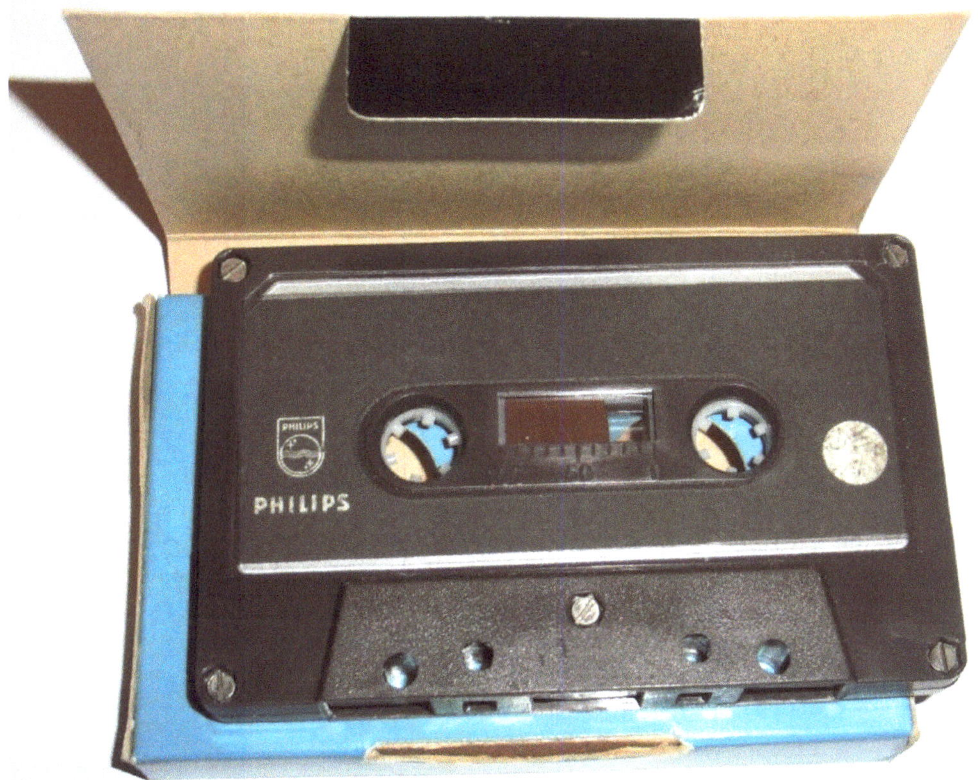

PHILIPS
PHILIPS
PHILIPS
MADE IN HOLLAND

INDEX

PHILIPS
MADE IN HOLLAND
EL 1903-01
PHILIPS

PHILIPS / NORELCO

1966

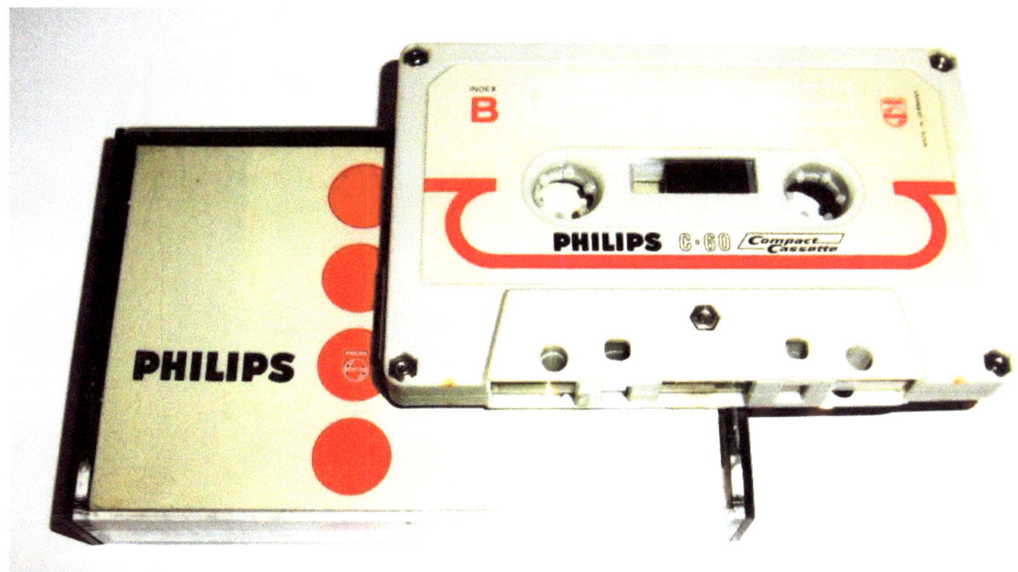

PHILIPS
C·60
Compact Cassette
PHILIPS C·60 Cassette mit 60 (2x30) Minuten Spielzeit.

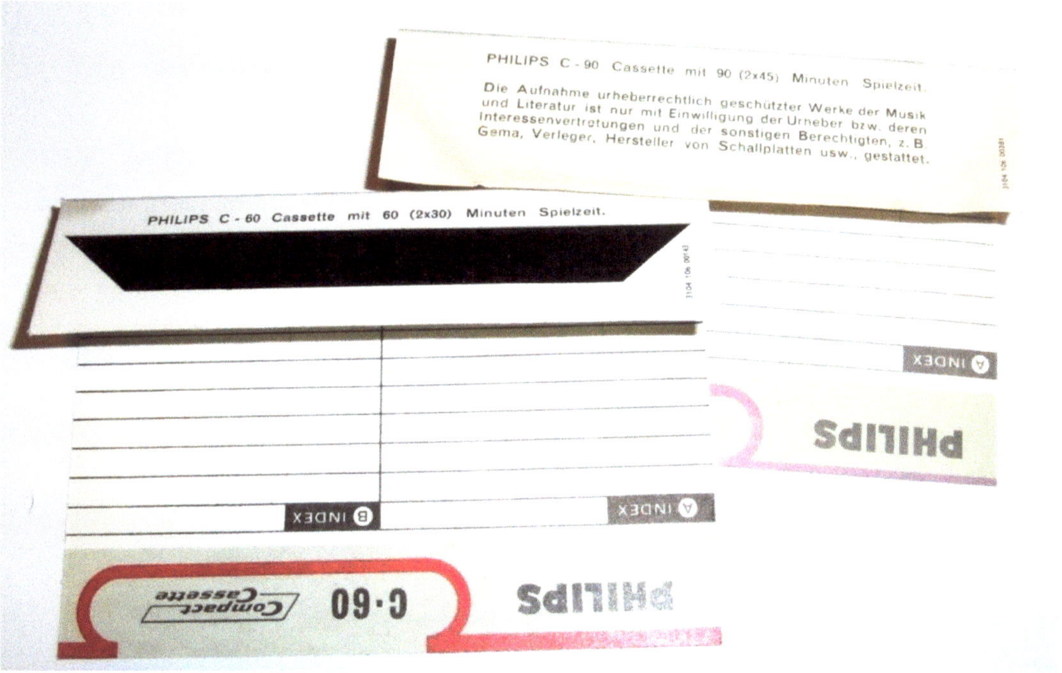

PHILIPS C - 90 Cassette mit 90 (2x45) Minuten Spielzeit.

Die Aufnahme urheberrechtlich geschützter Werke der Musik und Literatur ist nur mit Einwilligung der Urneber bzw. deren Interessenvertretungen und der sonstigen Berechtigten, z. B. Gema, Verleger, Hersteller von Schallplatten usw., gestattet.

PHILIPS C - 60 Cassette mit 60 (2x30) Minuten Spielzeit.

INDEX A

INDEX B

INDEX A

PHILIPS

PHILIPS

C·60

Compact Cassette

PHILIPS C-60 Cassette gives you 60 (2 x 30) minutes of playing time
PHILIPS C-60 Cassette bietet Ihnen 60 (2 x 30) Minuten Spielzeit
PHILIPS C-60 Cassette geeft U 60 (2 x 30) minuten speeltijd
La Cassette C-60 PHILIPS vous offre 60 (2 x 30) minutes d'enregistrement
El Chasis C-60 PHILIPS
PHILIPS C·60 Compact Cassette

PHILIPS
C·60 CASSETTE
PHILIPS
C·60 Compact Cassette

INDEX
A
INDEX
B
PHILIPS C·60 Compact Cassette

B
INDEX
PHILIPS
GERMANY

E IN GERMANY
PHILIPS
INDEX
B

PHILIPS
INDEX
B
PHILIPS C·90 Compact Cassette

PHILIPS C-90 Cassette gives you 90 (2x45) minutes of playing time
PHILIPS C-90 Cassette bietet Ihnen 90 (2x45) Minuten Spielzeit
PHILIPS C-90 Cassette geeft U 90 (2x45) minuten speeltijd
La Cassette C-90 PHILIPS vous offre 90 (2x45) minutes d'enregistrement
El Chasis C-90 PHILIPS le permite 90 (2x45) minutos de registro

3104 106 00273

PHILIPS C-90 Cassette mit 90 (2x45) Minuten Spielzeit.

Die Aufnahme urheberrechtlich geschützter Werke der Musik
und Literatur ist nur mit Einwilligung der Urneber bzw. deren
Interessenvertretungen und der sonstigen Berechtigten, z. B.
Gema, Verleger, Hersteller von Schallplatten usw., gestattet.

3104 106 00281

B INDEX   A INDEX

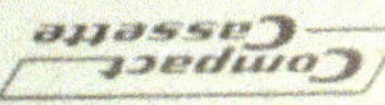

PHILIPS C-90 Compact Cassette
A
PHILIPS C-90 Compact Cassette
A
B
PHILIPS C-90 Compact Cassette
A
PHILIPS C-90 Compact Cassette

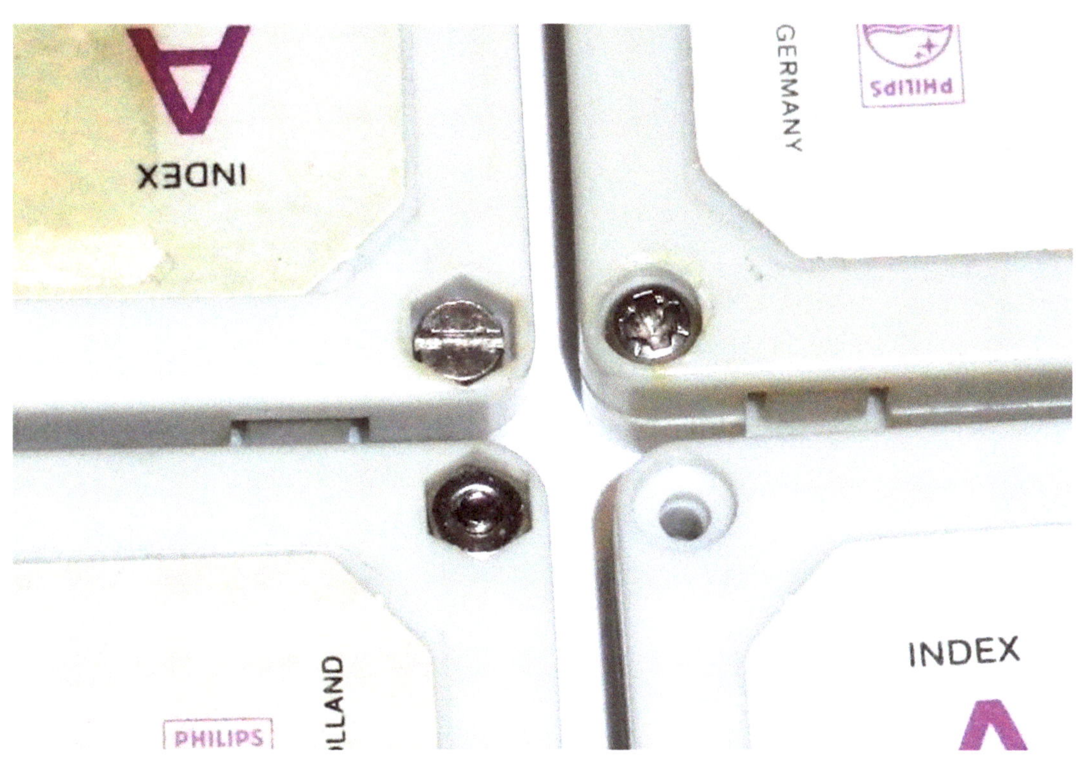
INDEX
GERMANY
PHILIPS
PHILIPS
HOLLAND
INDEX

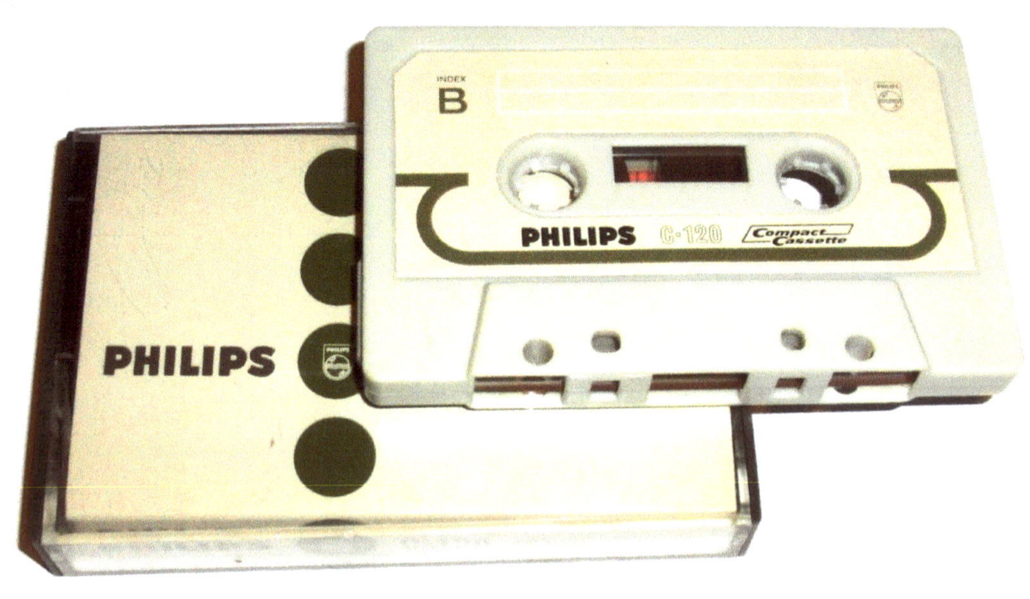
INDEX
B
PHILIPS C-120 Compact Cassette
PHILIPS

1971

PHILIPS
MADE IN HOLLAND
C-60
low noise
Compact Cassette

PHILIPS
MADE IN HOLLAND
1
C-60
low noise
Compact Cassette

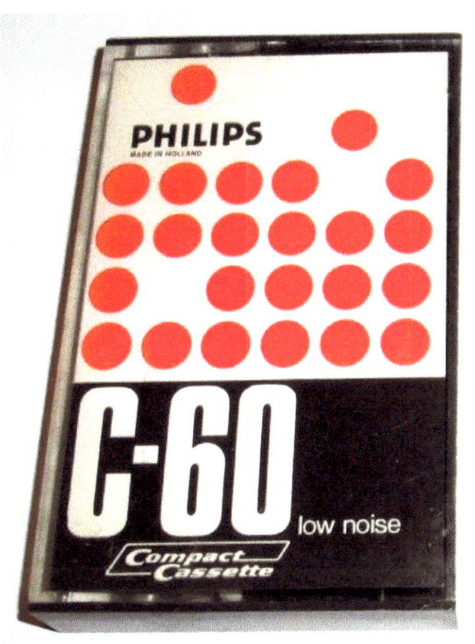

low noise
Compact Cassette
C-60
PHILIPS
1
low noise
Compact Cassette
C-60
PHILIPS
2
PHILIPS
MADE IN HOLLAND
C-60
low noise
Compact Cassette
1
PHILIPS
C-60
low noise
Compact Cassette
1

PHILIPS
index 2
index 1
PHILIPS

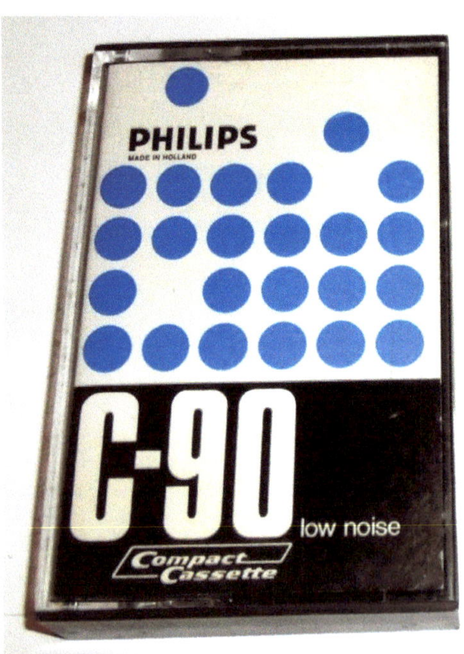

PHILIPS
MADE IN HOLLAND
C-90
low noise
Compact Cassette

PHILIPS
MADE IN HOLLAND
index
1
C-90
low noise
Compact Cassette

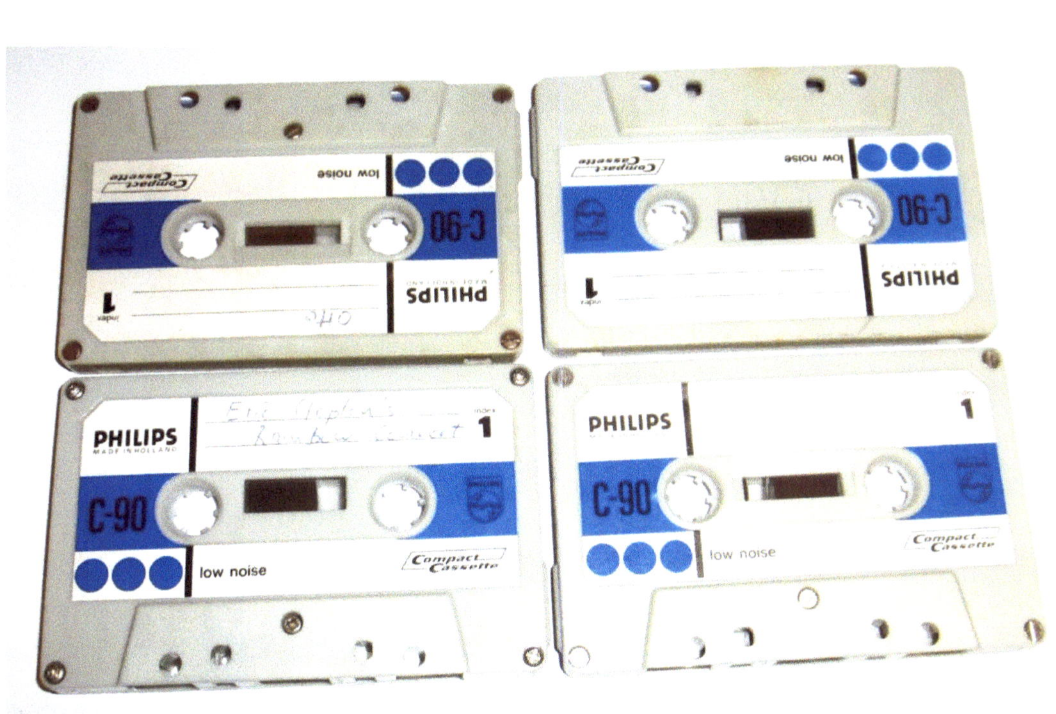

Compact Cassette
low noise
C-90
PHILIPS
MADE IN HOLLAND
PHILIPS
Compact Cassette
low noise
C-90
Eric Clapton's
Rainbow Concert
index 1
PHILIPS
C-90
low noise
Compact Cassette
index 1

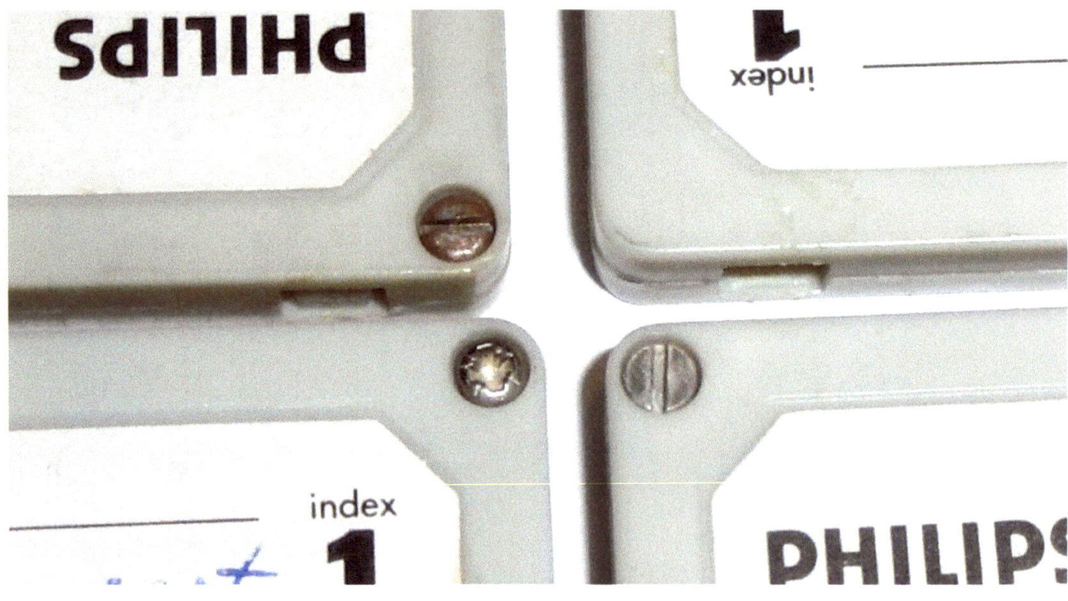

PHILIPS
index
index
1
PHILIPS

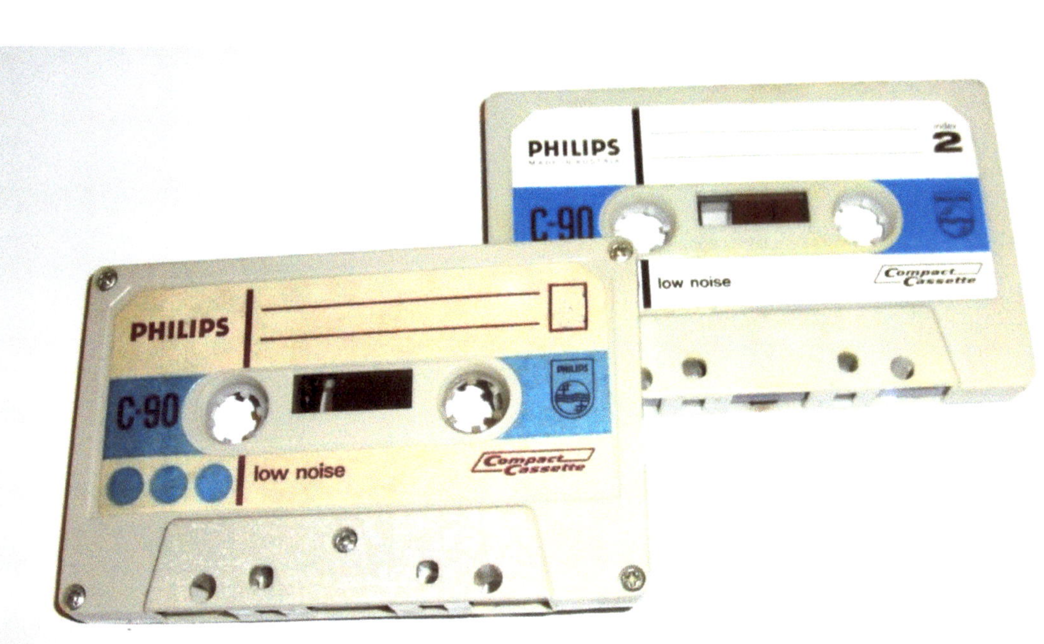
PHILIPS
2
C-90
low noise
Compact Cassette
PHILIPS
C-90
low noise
Compact Cassette

PHILIPS
C-120
low noise
Compact
Cassette

PHILIPS
MADE IN AUSTRIA
index
1
C-120
low noise
Compact
Cassette

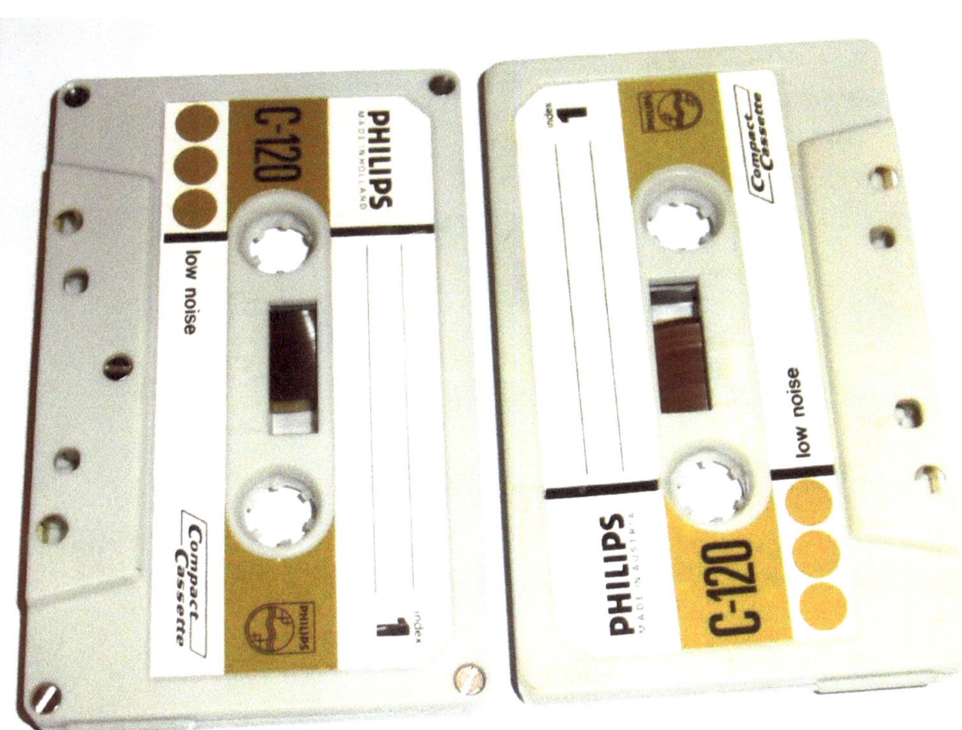

PHILIPS
MADE IN HOLLAND
C-120
low noise
Compact Cassette
index 1
PHILIPS
Compact Cassette
C-120
PHILIPS
MADE IN AUSTRIA
low noise
index 1

PHILIPS
MADE IN HOLLAND
CHROMIUMDIOXIDE
hi-fi Compact Cassette
C-60
HIGH FIDELITY INTERNATIONAL
hi fi
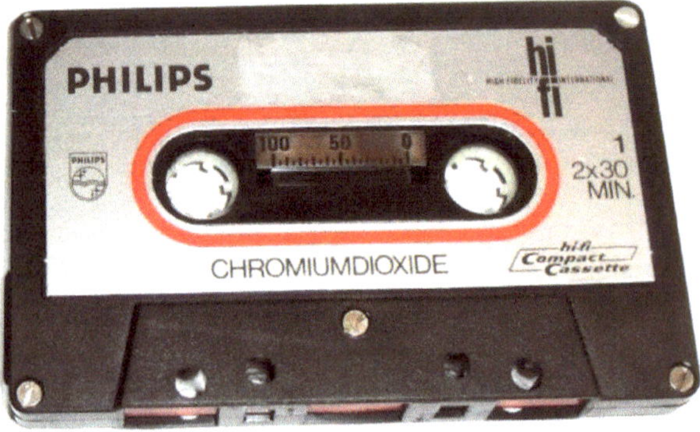
PHILIPS
hi fi
HIGH FIDELITY INTERNATIONAL
PHILIPS
100 50 0
1
2x30 MIN.
CHROMIUMDIOXIDE
hi-fi Compact Cassette

1975

PHILIPS
Made in Holland
STANDARD QUALITY
C-60
FERRO
LOW NOISE
ffloating
foil SECURITY

PHILIPS
Made in Holland
Compact Cassette
STANDARD QUALITY
C-60
2x30min
2
FERRIC
LOW NOISE
floating
foil SECURITY

PHILIPS
Made in Austria
STANDARD QUALITY
C-90
FERRO
LOW NOISE
ffloating
foil SECURITY

PHILIPS
Made in Austria
Compact Cassette
STANDARD QUALITY
C-90
2x45min
1
FERRO
LOW NOISE
ffloating
foil SECURITY

PHILIPS
Made in Austria
STANDARD QUALITY
C-120
FERRO
LOW NOISE
floating
foil SECURITY

PHILIPS
Made in Austria
Compact Cassette
STANDARD QUALITY
C-120
2x60min
1
FERRO
LOW NOISE
floating
foil SECURITY

PHILIPS
Made in Holland
SUPER QUALITY
C-60
HI-FERRO
LOW NOISE / H.O.
floating
foil SECURITY

PHILIPS
Made in Austria
Compact Cassette
SUPER QUALITY
C-60
2x30min
HI-FERRO
LOW NOISE / H.O.
floating
foil SECURITY
1

PHILIPS
Made in Austria
SUPER QUALITY
C-90
HI- FERRO
LOW NOISE / HO
floating
foil SECURITY

PHILIPS
Made in Austria
Compact Cassette
SUPER QUALITY
C-90
2x45min
100 50 0
1
HI FERRO
LOW NOISE / HO
floating
foil SECURITY

PHILIPS
Made in Holland
SUPER QUALITY
C-120
HI-FERRO
LOW NOISE / H.O.
floating
foil SECURITY

PHILIPS
Made in Holland
Compact Cassette
SUPER QUALITY
C-120
2x60min
1
HI-FERRO
LOW NOISE / H.O.
floating
foil SECURITY

PHILIPS
Made in Holland
HI-FI QUALITY
C-60
CHROMIUM
floating
foil SECURITY

PHILIPS
Made in Holland
hi-fi Compact Cassette
HI-FI QUALITY
C-60
2x30min
CHROMIUM
floating
foil SECURITY
1
hi fi

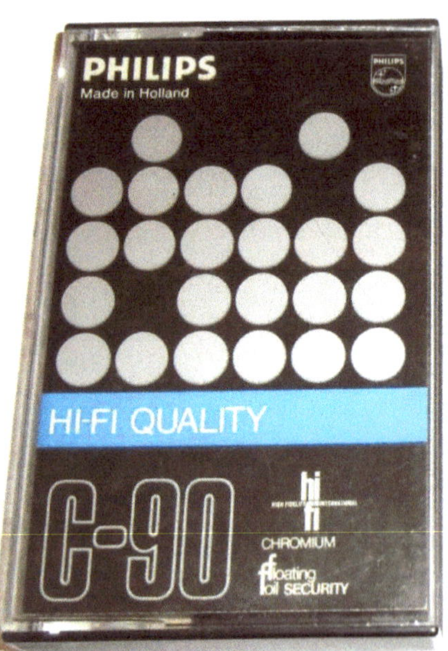
PHILIPS
Made in Holland
HI-FI QUALITY
C-90
CHROMIUM
floating foil SECURITY

PHILIPS
Made in Holland
hi-fi Compact Cassette
HI-FI QUALITY
C-90
2x45min
CHROMIUM
floating SECURITY
2

1978

PHILIPS
Made in Belgium
FERRO
LOW NOISE
SECURITY
C-60

FERRO
PHILIPS
Made in Belgium
C-60
Noise reduction

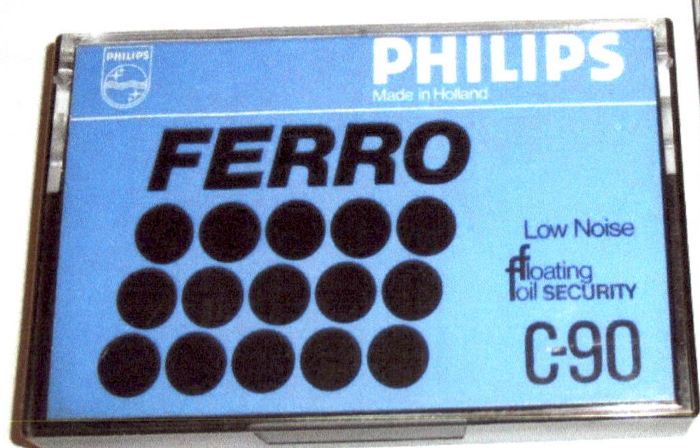
PHILIPS
Made in Holland
FERRO
Low Noise
floating
foil SECURITY
C-90

floating
foil security
FERRO
PHILIPS
Made in Holland
Compact Cassette
C-90
Low Noise
Noise reduction

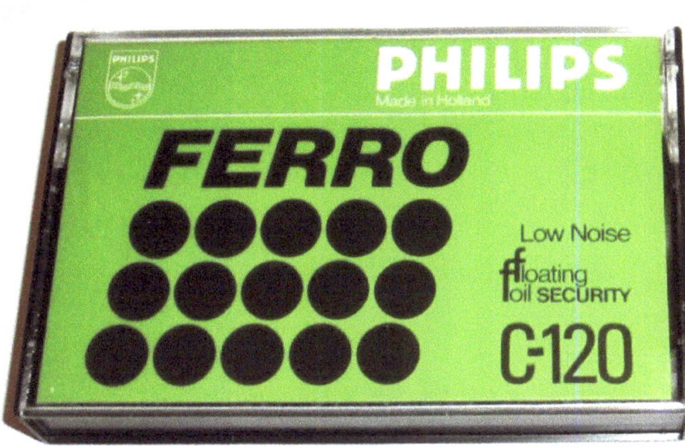

PHILIPS
Made in Holland
FERRO
Low Noise
floating
oil SECURITY
C-120

floating
oil security
FERRO
Compact Cassette
PHILIPS
C-120
Low Noise
Noise reduction

PHILIPS
Made in Holland
SUPER FERRO
High output LN
floating
foil SECURITY
C-60

floating
foil SECURITY
SUPER FERRO
Compact Cassette
PHILIPS
Made in Holland
C-60

PHILIPS
SUPER FERRO
High output LN
floating foil SECURITY
C-90

SUPER FERRO
PHILIPS

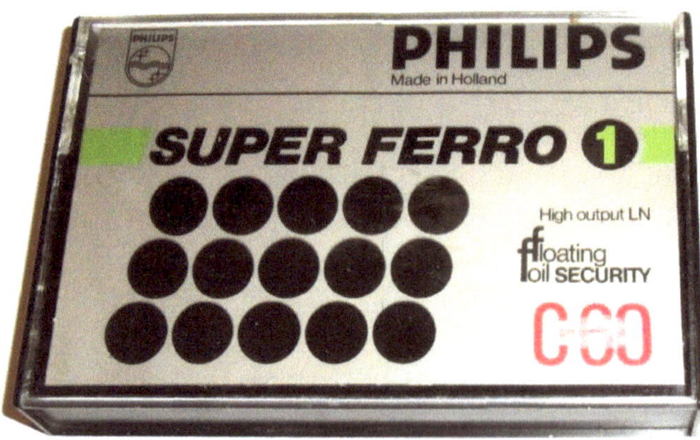
PHILIPS
Made in Holland
SUPER FERRO 1
High output LN
floating
oil SECURITY
C60

floating
oil security
SUPER FERRO 1
PHILIPS
Made in Holland
C-60
High output LN
Noise reduction

PHILIPS
Made in Holland
SUPER FERRO 1
PHILIPS
Made in Holland
SUPER FERRO 1
C-90
SECURITY
High output LH
Noise reduction

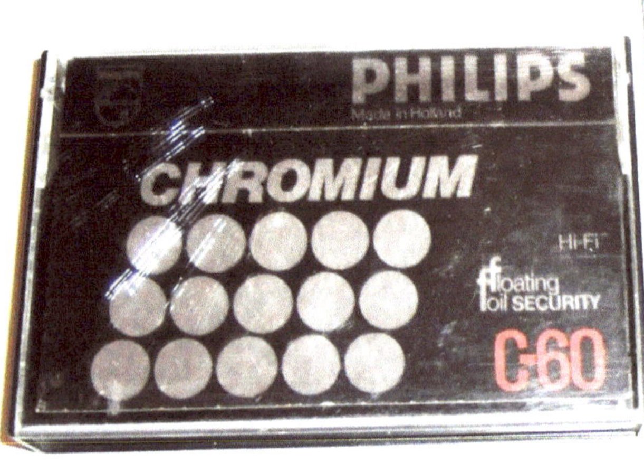
PHILIPS
Made in Holland
CHROMIUM
Hi-Fi
floating
foil SECURITY
C-60

floating
foil security
CHROMIUM
CHROMIUM
Hi-Fi
Compact Cassette
PHILIPS
Made in Holland
C-60
Hi-Fi

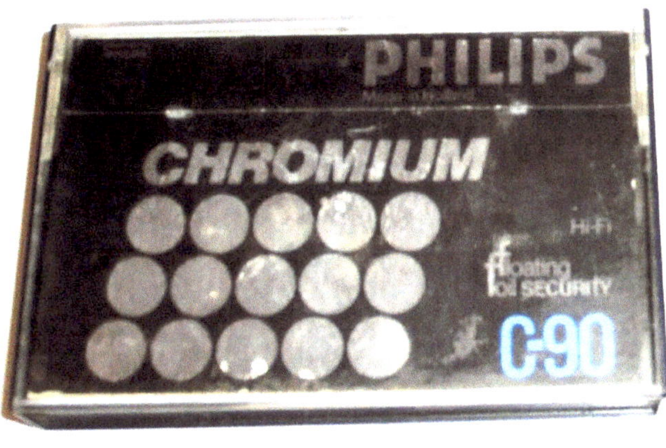

PHILIPS
CHROMIUM
Hi-Fi
floating foil SECURITY
C-90

floating foil security
CHROMIUM
CHROMIUM
PHILIPS
Made in Holland
Compact Cassette
C-90
HiFi

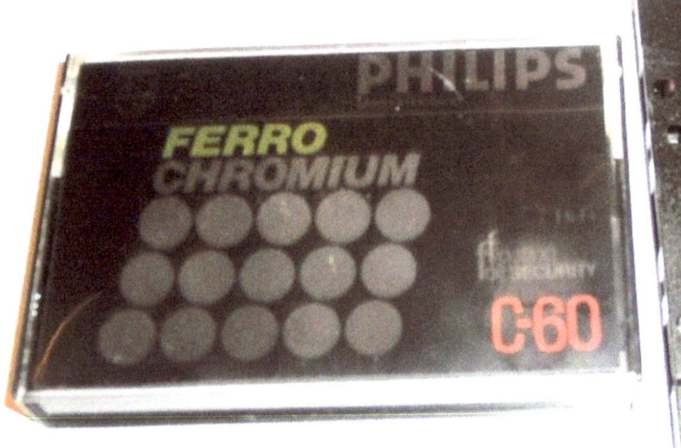

PHILIPS
FERRO
CHROMIUM
C-60

FERRO
CHROMIUM
PHILIPS
Made in Holland
C-60

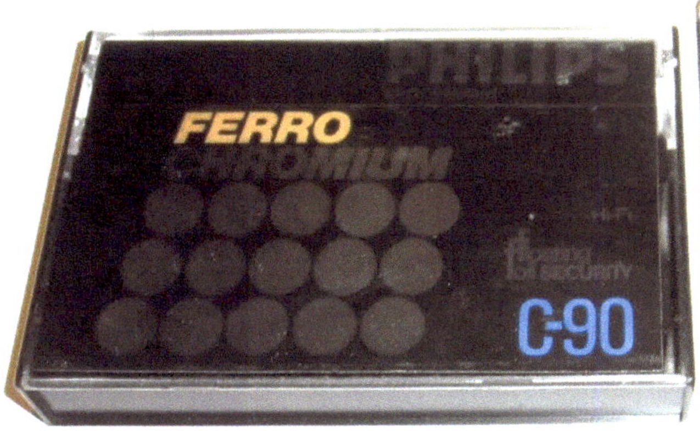

FERRO
CHROMIUM
HI-FI
C-90

FERRO
CHROMIUM
PHILIPS
Made in Holland
Hi-Fi
Compact Cassette
C-90
Hi-Fi

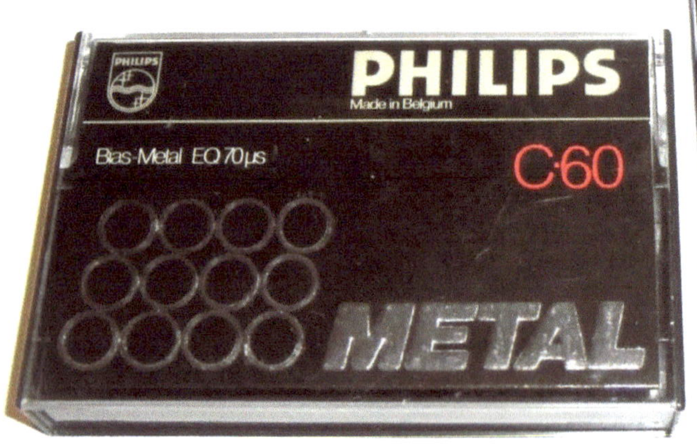
PHILIPS
Made in Belgium
Bias-Metal EQ 70μs
C-60
METAL

METAL
PHILIPS
Made in Belgium
Bias-Metal EQ 70μs
C-60
Noise Reduction

1978

PHILIPS
Made in Belgium
STUDIO QUALITY
FERRO
LOW NOISE
C-90
floating for SECURITY
PHILIPS
STUDIO QUALITY
PHILIPS
Made in Belgium
Compact Cassette
FERRO
LOW NOISE
C-90
Noise reduction

PHILIPS
Made in Belgium
floating
foil SECURITY
STUDIO QUALITY
FERRO
LOW NOISE
C-60

STUDIO QUALITY
PHILIPS
Compact Cassette
FERRO
LOW NOISE
C-60
Noise reduction

1981

PHILIPS
OF HOLLAND
StudioQuality C60
BIAS NORMAL EQUALISATION 120 µs
PHILIPS

PHILIPS
OF HOLLAND
SQ-C60
BIAS NORMAL EQUALISATION 120 µs
NOISE
REDUCTION
IN
OUT

PHILIPS
StudioQuality C90
BIAS NORMAL EQUALISATION 120 μs
MADE IN BELGIUM

PHILIPS
B side
INDEX
SO-C90
BIAS NORMAL EQUALISATION 120 μs
IN
OUT
NOISE REDUCTION

PHILIPS
OF HOLLAND
Chromium II
Studio Quality 60
TYPE II / BIAS: CrO₂ EQ 70µs
NOISE REDUCTION
IN
OUT
PHILIPS
OF HOLLAND
CSQ-90
TYPE II / BIAS: CrO₂ EQ 70µs

PHILIPS
OF HOLLAND
CSQ-60
TYPE II / BIAS: CrO₂ EQ 70µs
NOISE REDUCTION
IN
OUT

1981

PHILIPS
FFS
FLOATING FOIL
SECURITY
Compact Cassette
OUTPUT dB
FERRO
FERRO
FREQUENCY Hz
TYPE I / NORMAL BIAS, 120µs EQ
FERRO*C60
HIGH PRECISION MECHANISM
PHILIPS
FERRO
C60

PHILIPS
PHILIPS
FFS
FLOATING FOIL
SECURITY
OUTPUT dB
FERRO
FERRO
Compact Cassette
NORMAL BIAS. 120µs EQ
80 100 315 1000 3150 10K 20K FREQUENCY Hz
FERRO*C90
HIGH PRECISION MECHANISM

FFS
FLOATING FOIL
SECURITY
PHILIPS
NORMAL BIAS. 120µs EQ
HIGH PRECISION MECHANISM
FERRO
C90

PHILIPS
PHILIPS
ULTRA PRECISION MECHANISM
FFS
FLOATING FOIL
SECURITY
ULTRA
FERRO
C60
ULTRA
FERRO * C60
ULTRA PRECISION MECHANISM

PHILIPS
PHILIPS
FFS
FLOATING FOIL
SECURITY
ULTRA
FERRO
C90
ULTRA
FERRO * C90
ULTRA PRECISION MECHANISM

PHILIPS
MADE IN BELGIUM
Compact Cassette
ULTRA CHROME * C60
FFS
FLOATING FOIL
SECURITY
ULTRA PRECISION
MECHANISM
PHILIPS
ULTRA CHROME
C60
FFS
FLOATING FOIL
SECURITY
ULTRA PRECISION MECHANISM
INDEX

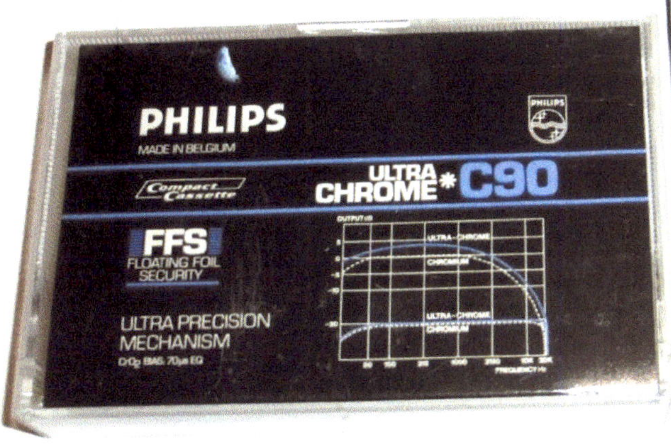

PHILIPS
MADE IN BELGIUM
Compact Cassette
ULTRA CHROME* C90
PHILIPS
FFS
FLOATING FOIL
SECURITY
ULTRA PRECISION
MECHANISM
CrO2 BIAS, 70µs EQ
OUTPUT dB
ULTRA - CHROME
CHROMIUM
ULTRA - CHROME
CHROMIUM
FREQUENCY Hz

FFS
FLOATING FOIL
SECURITY
PHILIPS
CrO2 BIAS, 70µs EQ
INDEX
ULTRA PRECISION MECHANISM
NOISE REDUCTION YES/NO
ULTRA CHROME
C90

PHILIPS
PHILIPS
FFS
FLOATING FOIL
SECURITY
METAL
C60
METAL*C60
AZIMUTH PRECISION CASSETTE MECHANISM

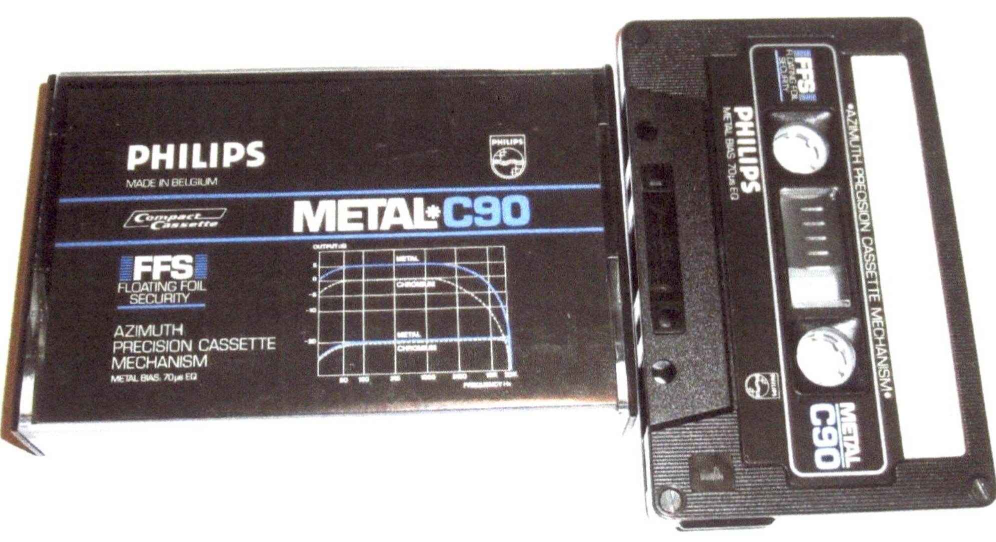
PHILIPS
MADE IN BELGIUM
Compact Cassette
METAL·C90
FFS
FLOATING FOIL
SECURITY
AZIMUTH
PRECISION CASSETTE
MECHANISM
METAL BIAS: 70µs EQ
PHILIPS
OUTPUT dB
METAL
CHROMIUM
METAL
CHROMIUM
FREQUENCY Hz
FFS
FLOATING FOIL
SECURITY
PHILIPS
METAL BIAS 70µs EQ
·AZIMUTH PRECISION CASSETTE MECHANISM·
METAL
C90
1

1984 - 1989

PHILIPS
PHILIPS
Ferro low noise
SQ*I
Type I Normal position 120µs Eq
60
Side A
SQ·I
60

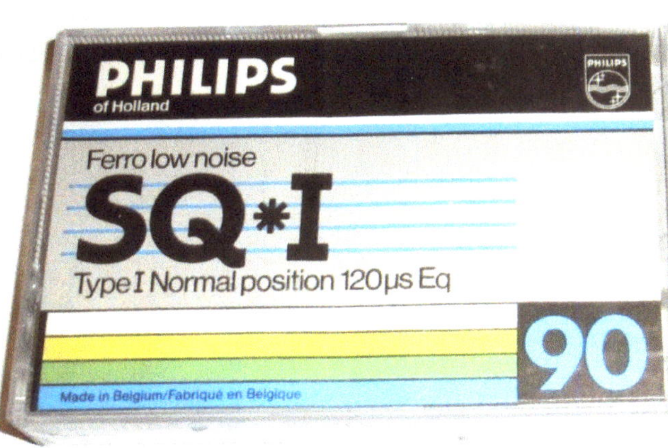
PHILIPS
of Holland
Ferro low noise
SQ*I
Type I Normal position 120μs Eq
90
Made in Belgium/Fabriqué en Belgique

PHILIPS
Side B
Ferro low noise
Type I Normal position 120μs Eq
SQ*I
90
2

FE∗I 60
PHILIPS
MADE IN BELGIUM
FABRIQUE EN BELGIQUE
Ferro tape quality
High performance
Low distortion
Type I, normal position - 120 µs EQ
Philips Ferro Band
Hohe Aussteuerfähigkeit
Geringe Verzerrung
Type I, Position „normal" - 120µs EQ
Qualité de bande Ferro
Haute Performance
Distortion très faible
Type I, polarisation: „normale" - 120µs EQ
Ferro tape kwaliteit
Hoge uitstuurbaarheid
Lage vervorming
Type I, Positie „normal" - 120µs EQ
OUTPUT dB
FREQUENCY Hz
TAPE SELECTOR
IEC POSITION BIAS EQ
I Normal Normal 120µs
REC. TIME
60 min.
(2x30)
LENGTH
90 m.
PHILIPS
FE∗I 60
TYPE I NORMAL POSITION - 120µs EQ
N.R.
YES
NO

PHILIPS
A
B
COUNT
COUNT
3104 106 61841
NR   YES   NO
PHILIPS   FE·I   90
TYPE I·NORMAL BIAS·120µs EQ

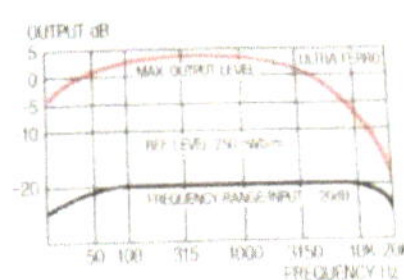

TAPE SELECTOR				REC. TIME	LENGTH
IEC	POSITION	BIAS	EQ	60 min. (2x30)	90 m
I	Normal	Normal	120µs		

PHILIPS
UC*II 90
New improved Chrome Tape
Type II, high position - CrO₂ - 70 μs EQ
High dynamics
Ultra precision mechanism
Nouvelle qualité de bande Ultra Chrome
Position: CrO₂ - 70 μs EQ Type II
Haut niveau de sortie et dynamique
performante
Mécanisme de très haute précision
Verbesserte Chrom Kassettenband
Type II, position „high" CrO₂ - 70 μs EQ
Hoge uitstuurbaarheid
Ultra precisie mechanisme
Nieuws verbessertes Ultra Chrome
Qualitätsband
Type II Position „high" CrO₂ - 70 μs EQ
Mit grossem Dynamik Bereich
Ultra Präzisions Gehäuse
TAPE SELECTOR
TEC POSITION BIAS EQ
II High High 70μs
REC. TIME
90 min.
(2x45)
LENGTH
135 m.
PHILIPS
PHILIPS
UC-II 90
TAPE II HIGH POSITION - 70μs EQ

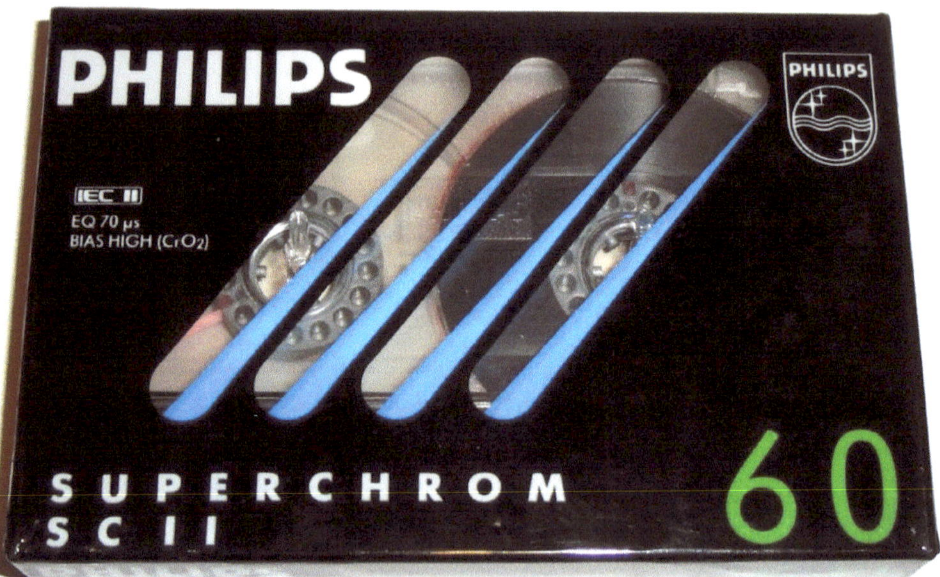

PHILIPS
PHILIPS
IEC II
EQ 70 µs
BIAS HIGH (CrO2)
SUPERCHROM
SC II
60

PHILIPS
PHILIPS
S C II
90
2x
45 MIN.

Beste Qualität und Zuverlässigkeit dank computer-
gesteuerter Fabrikation
Magnetband mit feinster Chromdioxydbeschichtung
für aussergewöhnlich klare Wiedergabe
Hohe Dynamik über den gesamten Frequenzbereich
Volle Ausnützung der Dynamik des Bandmaterials
durch patentierte Wickelkerne
Hochpräzises Gehäuse, welches die IEC-Anforde-
rungen bei weitem übertrifft

Top quality and reliability thanks to computer-
controlled production
Magnetic tape with superfine chrome dioxide coating
for exceptionally clear reproduction
Excellent dynamic quality over the entire frequency
range
High-precision shell which surpasses the IEC
requirements by far

Qualité et fiabilité de premier ordre, grâce à la
fabrication commandée par ordinateur
Bande magnétique à revêtement ultra-fin d'oxyde de
chrome, assurant une reproduction extraordinairement claire
Ample plage dynamique dans l'ensemble de l'intervalle
de fréquences
Exploitation intégrale de la plage dynamique du matériau
de la bande, grâce aux axes de bobinage brevetés
Boîtier de haute précision, dépassant largement les
exigences CEI

Il meglio della qualità e dell'affidabilità, grazie alla
fabbricazione comandata da elaboratore
Nastro magnetico con finissimo rivestimento di ossido
cromo, per una riproduzione straordinariamente chiara
Ampio intervallo dinamico nell'intero campo di frequenza
con i valori più favorevoli nella zona del rumore di banda
Scatola di alta precisione, ampiamente superiore alle
esigenze CEI

FREQUENCY RESPONSE
Level dB
Frequency Hz

IEC II
EQ 70 µs
BIAS HIGH (CrO2)

7 610783 000004

PHILIPS

PHILIPS
EQ*I
REC. TIME 90min. (2x45)
LENGTH 135m.
90
PHILIPS
MADE IN BELGIUM-
FABRIQUE EN BELGIQUE
TYPE I - NORMAL POSITION
120µs EQ
PHILIPS
90
TYPE I NORMAL POSITION
120µs EQ
EQ·I 90
INDEX
N.R.
On
Off
2

PHILIPS
PHILIPS
N.R.(   )
YES☐ NO☐
60
NORMAL POSITION
FS
MOVING SOUND
PHILIPS
N.R.(   )
YES☐ NO☐
60
FS
TYPE I·NORMAL POSITION
MOVING SOUND
PHILIPS
N.R.(   )
YES☐ NO☐
TYPE I·NORMAL
MOVING SOUND

A
ON
DATE/TIME
NOISE REDUCTION
B
PHILIPS
TYPE I · NORMAL POSITION 120 µs EQ
FS 60
B
PHILIPS
PHILIPS
FS
TYPE I · NORMAL POSITION 120 µs EQ
90

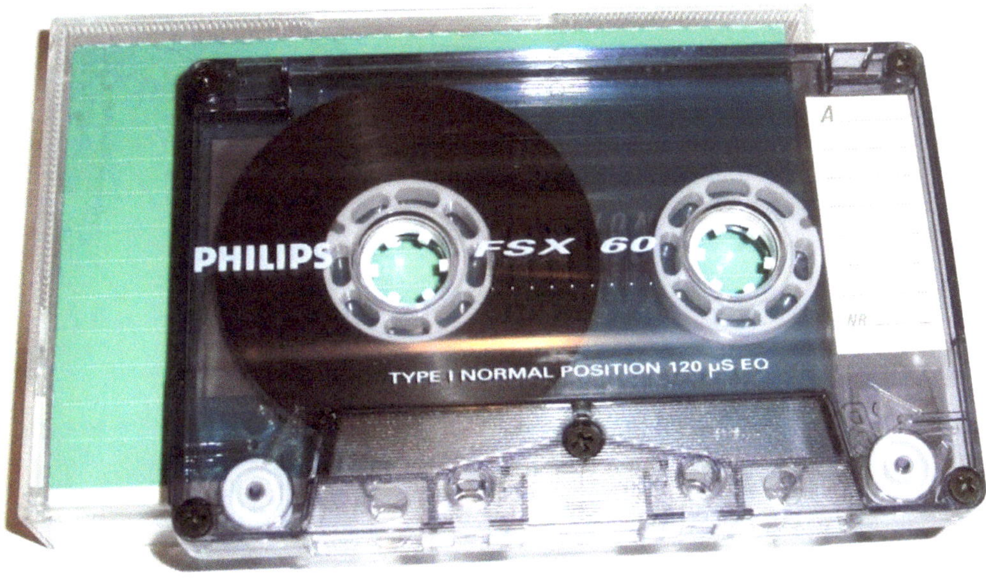
PHILIPS FSX 60
TYPE I NORMAL POSITION 120 µS EQ
A
NR

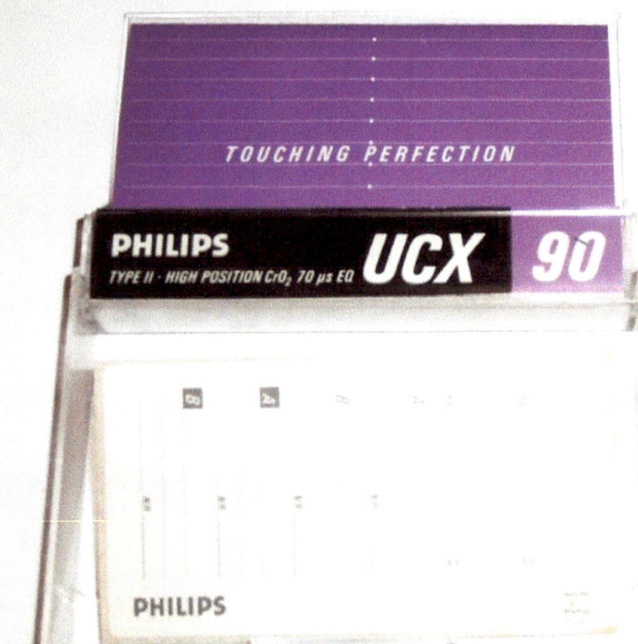

TOUCHING PERFECTION
PHILIPS
UCX 90
TYPE II · HIGH POSITION CrO₂ 70 µs EQ
PHILIPS

PHILIPS
UCX · 90
TYPE II HIGH POSITION CRO₂ 70µS EQ

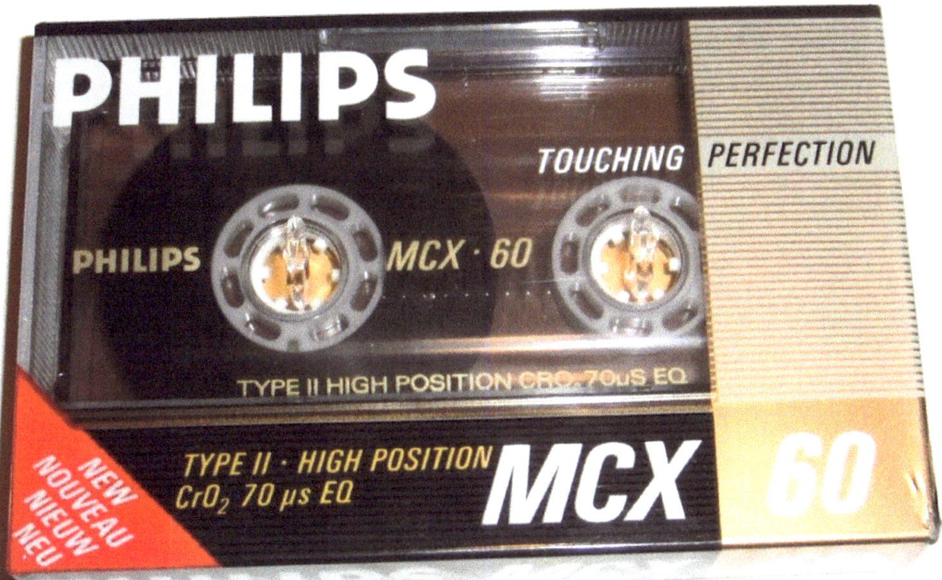
PHILIPS
TOUCHING PERFECTION
PHILIPS MCX · 60
TYPE II HIGH POSITION CRO₂ 70uS EQ
NEW
NOUVEAU
NIEUW
NEU
TYPE II · HIGH POSITION
CrO₂ 70 µs EQ
MCX 60

Normal Position (Type I) 120μs
FERRO
PHILIPS
FS 60
PHILIPS
FS
PHILIPS
FS
TYPE I NORMAL POSITION 120 μS EQ
PHILIPS
FERRO
FS 90
Normal Position (Type I) 120μsEQ

Normal Position (Type I) 120µsEQ
PHILIPS
FERRO
FSX 90
PHILIPS
FSX

1990

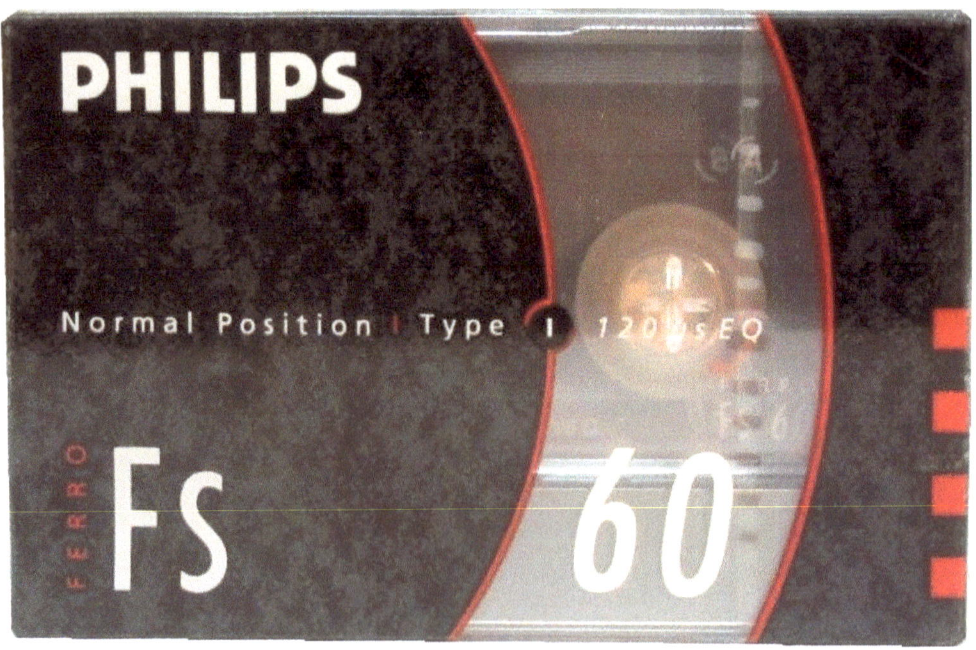

PHILIPS
Normal Position | Type I | 120 µs EQ
FERRO
FS
90
PHILIPS
FS 90

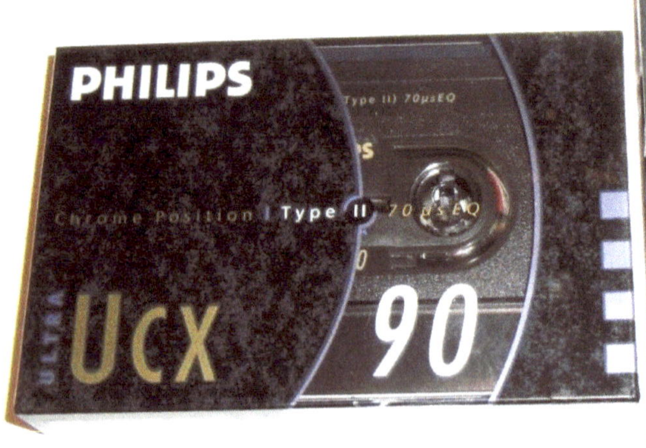
PHILIPS
Chrome Position I Type II 70µsEQ
ULTRA
UCX
90

PHILIPS
UCX 60
ULTRA
Chrome Position (Type II) 70µsEQ
A

PHILIPS
FERRO FS
Normal Position I Type I
90
PHILIPS
Metal Position
PLUS METAL
100
PHILIPS
Chrome
CD EXTRA
60
PHILIPS

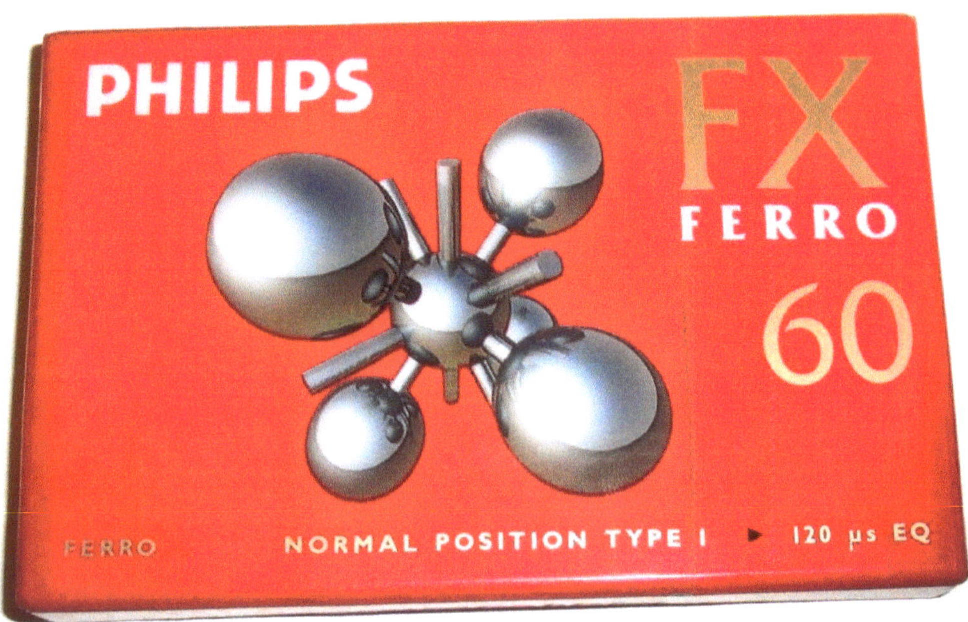

1994

PHILIPS
CD
ONE
60
NORMAL POSITION TYPE I   120 µs EQ
CD
ONE
60

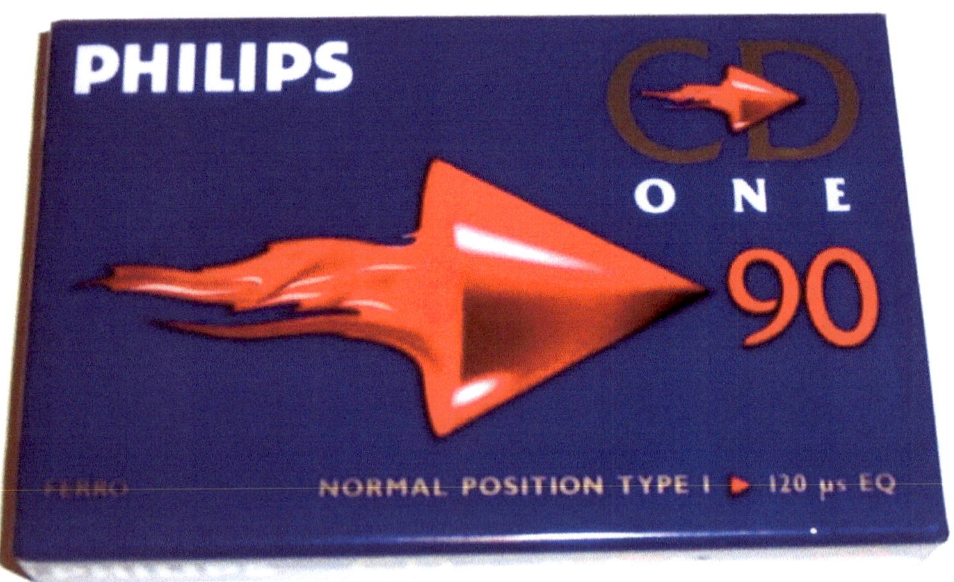
PHILIPS
CD
ONE
90
FERRO
NORMAL POSITION TYPE I   120 µs EQ

PHILIPS
CD PLUS
60
SUPERFERRO
NORMAL POSITION TYPE I 120 µs EQ

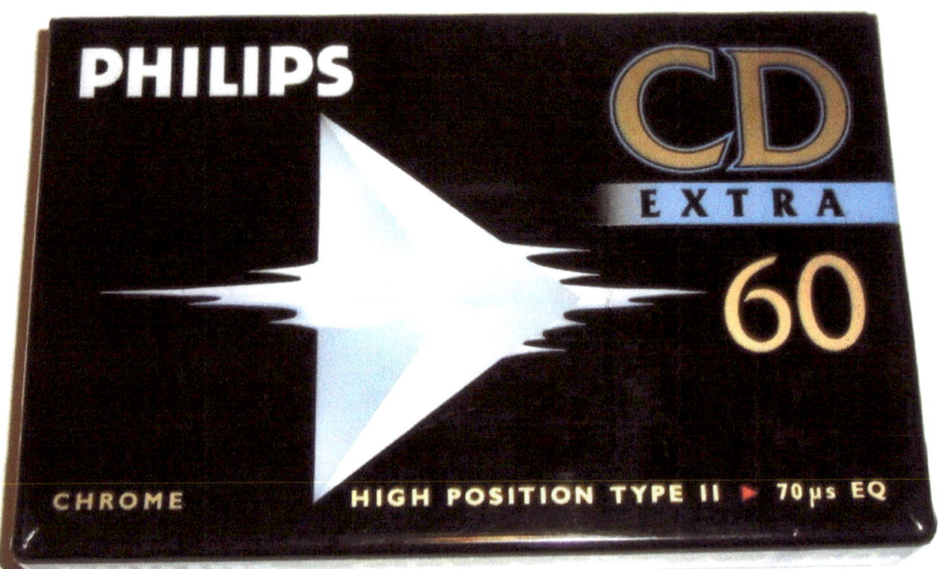

PHILIPS
CD
EXTRA
60
CHROME
HIGH POSITION TYPE II ▶ 70 µs EQ

PHILIPS
CD
EXTRA
90
CHROME
HIGH POSITION TYPE II ▶ 70 µs EQ

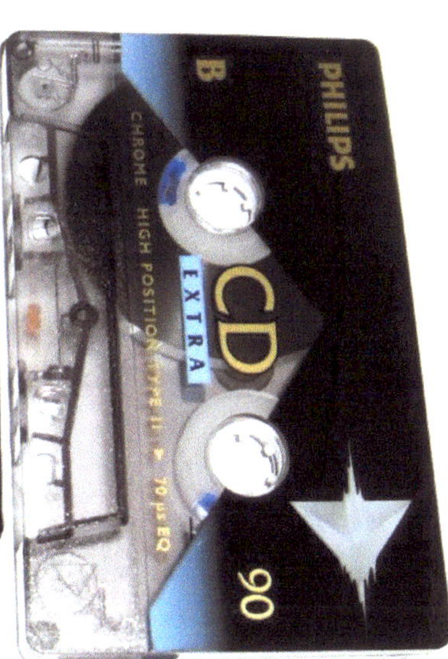

B
PHILIPS
CHROME HIGH POSITION TYPE II ▶ 70 µs EQ
CD
EXTRA
90

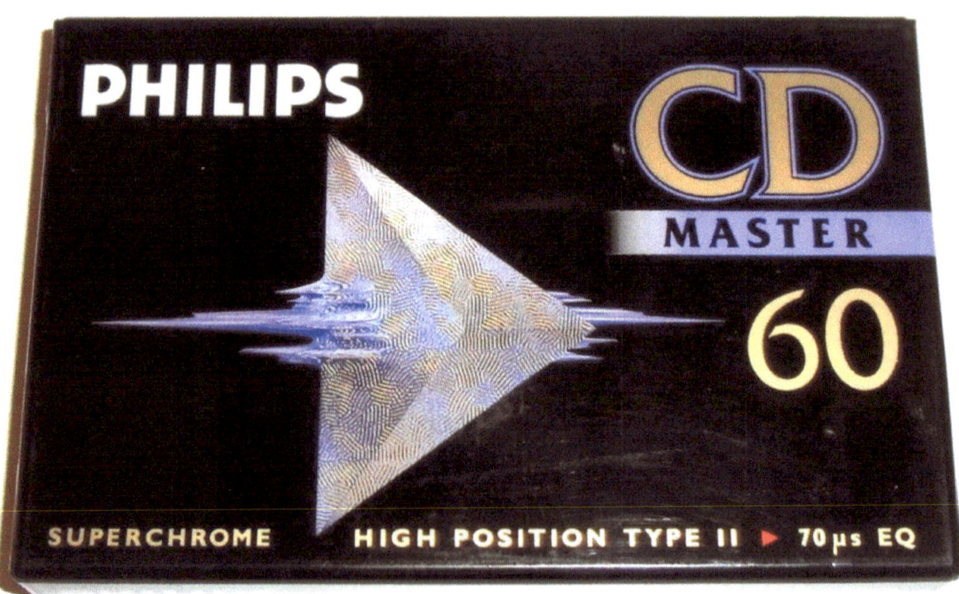
PHILIPS
CD
MASTER
60
SUPERCHROME
HIGH POSITION TYPE II
70 µs EQ

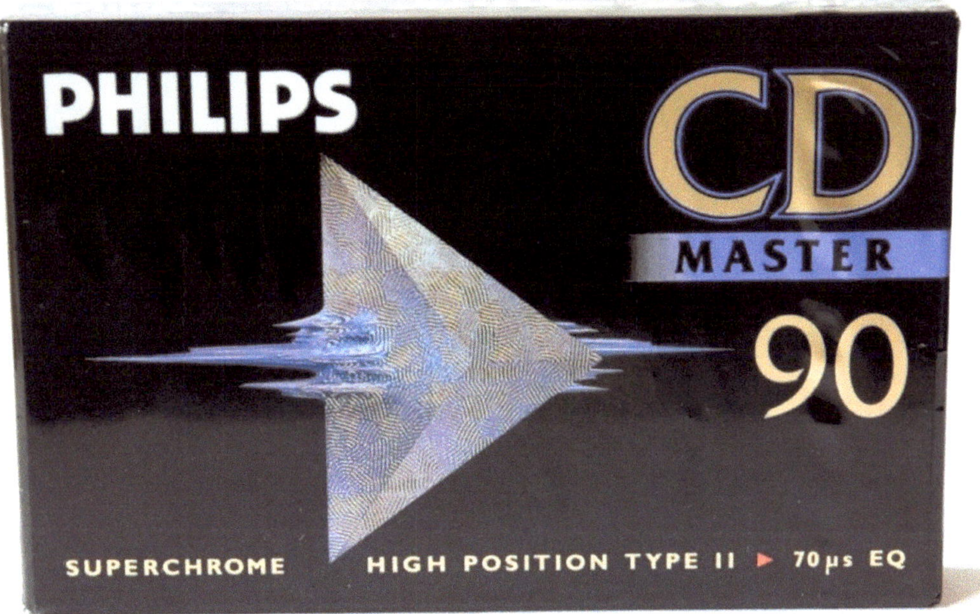

PHILIPS
CD
MASTER
90
SUPERCHROME
HIGH POSITION TYPE II ▶ 70 µs EQ

1997 - 1999

PHILIPS
CD one 60
120 µs EQ   type I   ferro position
PHILIPS
CD one 90
ferro position

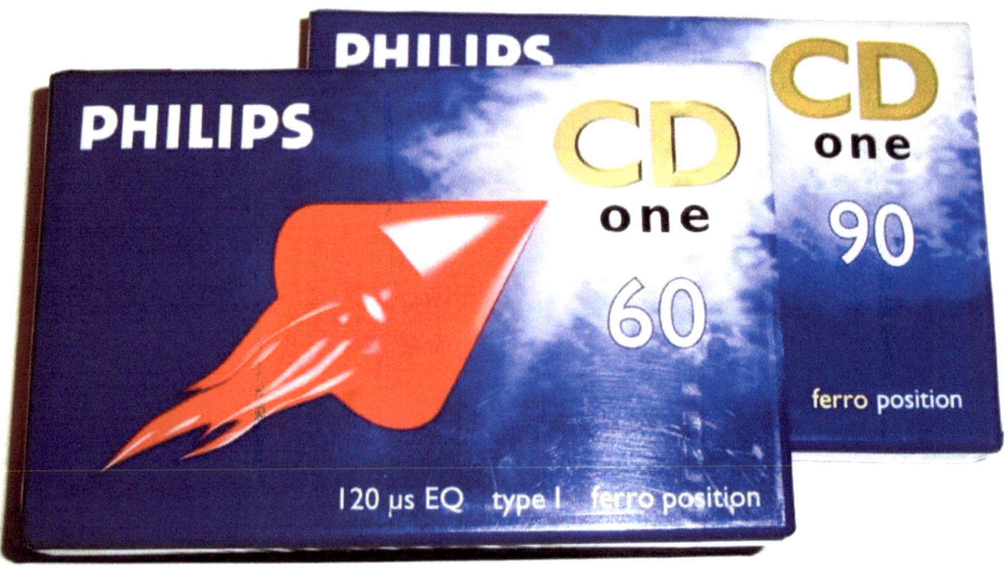

PHILIPS
CD plus
90
PHILIPS
CD plus
60
ferro+
120 µs EQ   type I   ferro position
ferro position

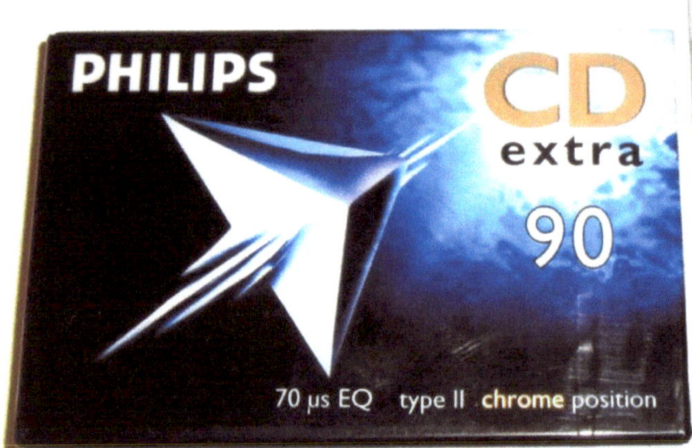

PHILIPS
CD
extra
90
70 µs EQ   type II   chrome position

A
TYPE II
PHILIPS
CHROME POSITION
70 µs EQ
CD
extra
60

2002

PHILIPS
FX
FERRO
60
FERRO
NORMAL POSITION
PHILIPS
FX
FERRO
A
FERRO
NORMAL POSITION TYPE I

Add to Compact Cassette Recorder

Add to Compact Cassette Recorder

Add to Compact Cassette Recorder

NORELCO

NORELCO TAPE
CARTRIDGE
FOR USE WITH CARRY-CORDER®"150'
Norelco® TAPE CARTRIDGE
Norelco® C-120
Compact Cassette
LIFETIME GUARANTEE

Norelco®
C-120
Compact Cassette
LIFETIME GUARANTEE

Norelco® C-120
LIFETIME GUARANTEE

INDEX
SIDE
A
100   50   0
Norelco
MADE IN HOLLAND
Compact
Cassette
C-90
A INDEX   D
Norelco
Compact
Cassette

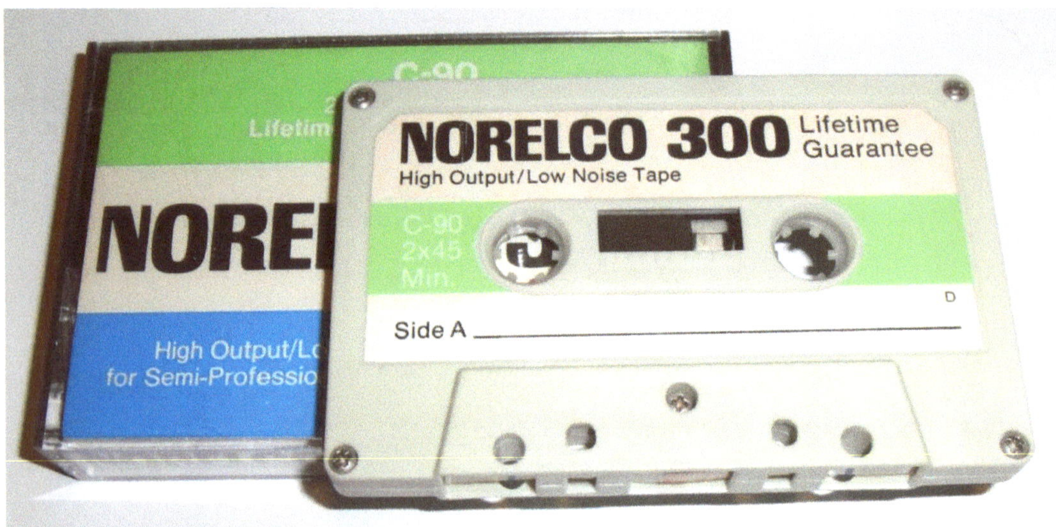
NORELCO 300
Lifetime Guarantee
High Output/Low Noise Tape
C-90
2x45
Min.
Side A
D
NORE
Lifetime
NORE
High Output/Lo
for Semi-Professio

FROM:
TO:
FIRST CLASS 20¢
PLACE
POSTAGE
HERE
THIRD CLASS 8¢
Norelco

Product of PHILIPS

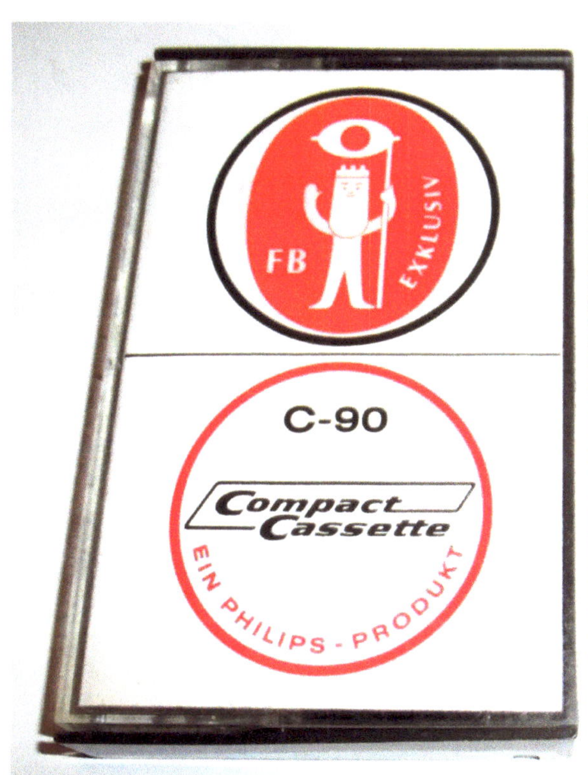

Product of PHILIPS

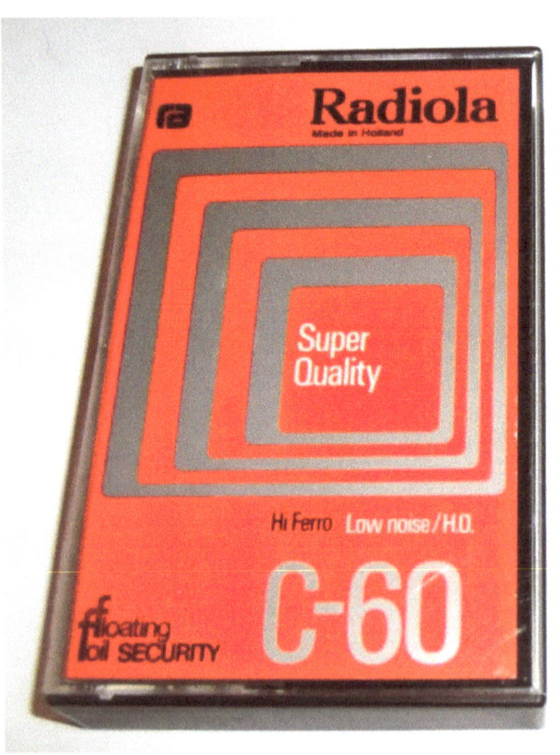

Product of PHILIPS

Product of PHILIPS

Product of PHILIPS

Product of PHILIPS

Product of PHILIPS

PHILIPS EINLOCHKASSETTE - Type: Normal
wurde nie der Öffentlichkeit vorgestellt, in Wien produziert

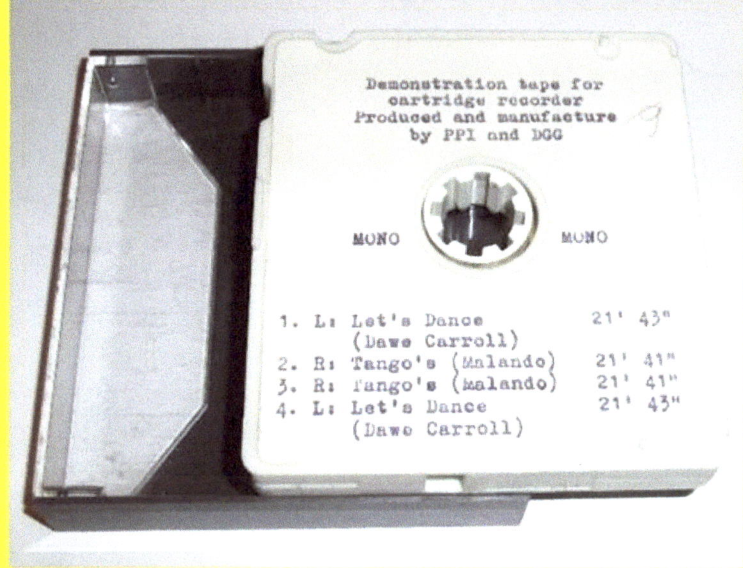

Demonstration tape for
cartridge recorder
Produced and manufacture
by PPI and DGG

MONO   MONO

1. L: Let's Dance   21' 43"
   (Dawe Carroll)
2. R: Tango's (Malando)   21' 41"
3. R: Tango's (Malando)   21' 41"
4. L: Let's Dance   21' 43"
   (Dawe Carroll)

Uwe H. Sültz - Compact Cassetten Bücher

PHILIPS EINLOCHKASSETTE - Type: Normal
PHILIPS entschied sich für die vom Team Lou Ottens entwickelte
Zweilochkassette, die zukünftig Compact Cassette genannt wurde

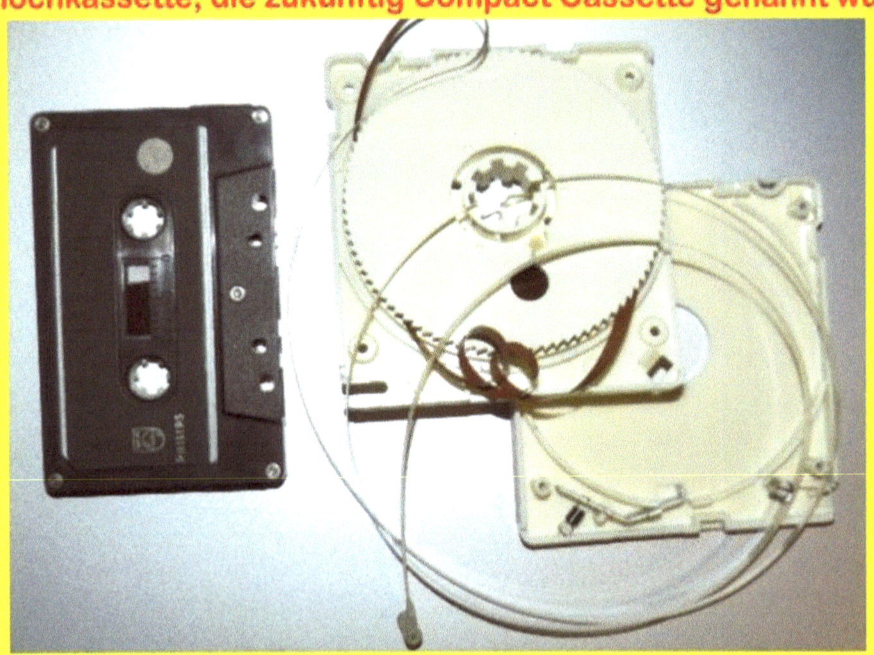

Uwe H. Sültz - Compact Cassetten Bücher

PHILIPS COMPACT CASSETTE EL 1903 - Type: Normal - Jahr: 1963

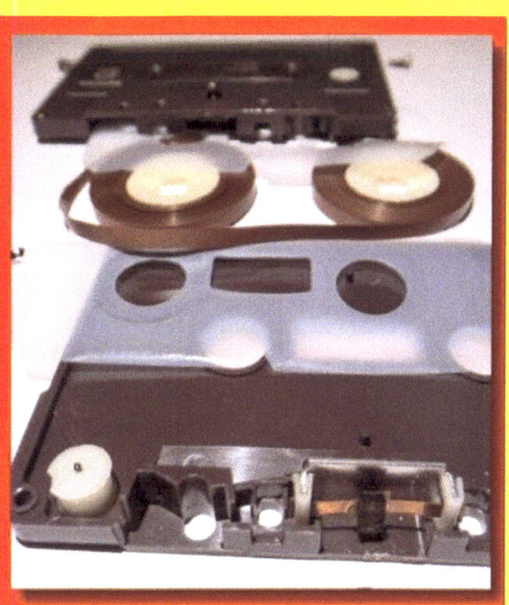

Uwe H. Sültz - Compact Cassetten Bücher

PDM Compact Cassetten - Type: Chrom + Metal - Jahr: 1981

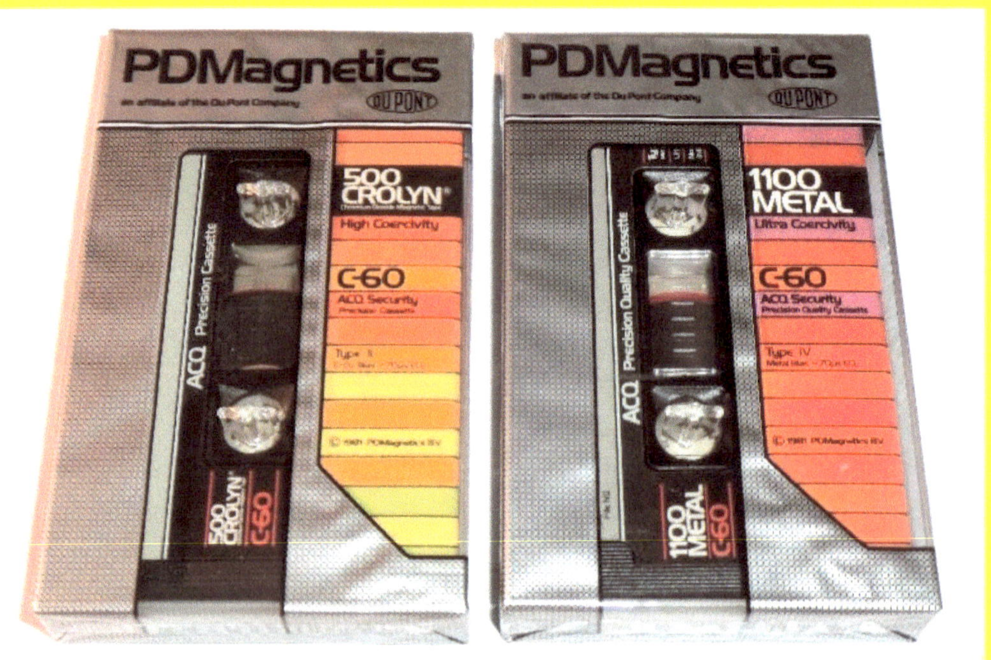

PDMagnetics
an affiliate of the Du Pont Company
DU PONT
500 CROLYN
High Coercivity
C-60
ACQ Security
Precision Cassette
Type II
ACQ Precision Cassette
500 CROLYN C-60
PDMagnetics
an affiliate of the Du Pont Company
DU PONT
1100 METAL
Ultra Coercivity
C-60
ACQ Security
Precision Quality Cassette
Type IV
ACQ Precision Quality Cassette
1100 METAL C-60

Uwe H. Sültz - Compact Cassetten Bücher

PDM Compact Cassette - Type: Chrom - Jahr: 1990
PDM
CHROME POSITION
CD 90
IEC TYPE II
HIGH POSITION
Super precision cassette casing.
Improved temperature resistance.
Wide dynamic range.
Minimum de distor
Resiste aux tempe
Meilleure précisio
fréquences.
Elimina ruidos anomalos.
Cassetes de "super precisão"
resistente a temperaturas elevadas.
Precisão perfeita em todas as
frequências.
Geruschlose Band
low noise.
Hohe Temperaturb
Breites dynamische
HIGH BIAS EQ 70 µS
PDM
CD 90
FE
FEI
CD
CDI
Uwe H. Sültz - Compact Cassetten Bücher

Uwe H. Sültz - Compact Cassetten Bücher

Keine Garantie auf Vollständigkeit!

No guarantee of completeness!